Note: the photos used are for illustration purposes only and may not reflect the species listed.

Ordering information: **LeaJacobson.com/SafetyMadeSimple**

A Message from the Author

If you're reading this book, you know the value of using essential oils safely - *thank you*. We need more of you, as the popularity of essential oils has grown exponentially since I first started learning about them almost a dozen years ago.

Another thing that's grown? Unsafe information.

Anyone can join a company and start posting their opinions about essential oils all over social media - and they do.

This is how so many people find my website: they've followed the widely-circulated advise, gotten hurt, and they're searching for answers.

Hopefully you're reading this before you experience any of the potential adverse reactions that can happen when people use essential oils before learning about them.

But if you're one of the countless people who have gotten hurt, I'm sorry. I'm doing my best to bring safety awareness to the essential oil world *before* more people get hurt, and I hope this book can serve as a guide to remaining safe from this day forward.

Keep in mind *I don't sell essential oils*. And these safety guidelines apply to *all brands*.

Remember, "Safety is safety, no matter the brand!"

If you have any questions, feel free to email me at Lea@UsingEOsSafely.com

START HERE

What Makes This Book Different

Essential Oil Safety Made Simple takes the guesswork out of the common question, "is this essential oil safe?" Using the color-coded **Essential Oil Safety Classification System™** I created, you can see at a glance if the essential oil is safe to use, if caution is needed, or to avoid altogether.

This makes essential oil safety...simple!

I've also included specific ages, topical maximums, drug interactions, and other contraindications and interactions where applicable.

The classifications (for HUMANS) are as follows:

SAFE for ALL

GREEN = **SAFE for ALL HUMANS* when general safety guidelines are adhered to.** There are *no known issues* with these essential oils, although rare allergies may exist.

<u>General Safety Guidelines for Inhalation</u>: Avoid using with babies under 6 months old. For adults, diffuse no more than 10 minutes at a time with a 20 minute break, as needed.

<u>General Safety Guidelines for Topical Use</u>: Avoid applying on children under 2 years old unless with intention (bug bite, burn, etc). For adults, always dilute before applying to the skin, using a dilution appropriate for the health issue (1-5% for most issues).

<u>Ingestion</u>: don't do it. Ingestion contraindications and interactions are included for reference only.

*May not be safe for dogs.

CAUTION for SOME

YELLOW = **USE CAUTION! Not safe for everyone, and/or has a topical max.** There are safety issue(s) to consider, or a topical max to adhere to, so you can use this essential oil and avoid adverse reactions.

CAUTION for ALL

RED = **STOP! Safety issues affecting everyone.** Safety issues are unavoidable if you use this essential oil. You may decide to use these anyways, if benefit outweighs risk, but most people should avoid these essential oils.

DO NOT USE

BLACK = **DO NOT USE - unsafe for all.** There is no reason to risk using this essential oil.

In addition to color-coded headers, all known safety issues are provided for clarity and full disclosure. This information is reflected by **easy-to-reference Safety Icons**, as well as full descriptions so there is no confusion. With this book in your possession, you can be confident that you'll be using essential oils safely!

Safety Icon Legend

SAFETY BY AGE

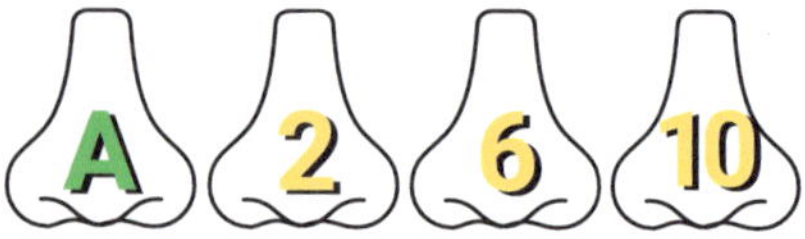

Indicates **inhalation** with the **minimum age in years**:
2 = 2+ years, **6** = 6+ years, **10** = 10+ years, **A** = "all" although in general **avoid around children under six months old** and always follow safe inhalation guidelines.

Not safe for any age to inhale.

Indicates **topical use** with the **minimum age in years**:
2 = 2+ years, **6** = 6+ years, **10** = 10+ years, **A** = "all" although in general **avoid using on children under two years old** and always follow safe dilution guidelines.

Not safe for any age to use topically.

PREGNANCY SAFETY

 Indicates the essential oil **IS SAFE** to use during **pregnancy** and **breastfeeding**.

 Indicates the essential oil IS **NOT recommended to use** during pregnancy and breastfeeding.

Pregnancy note: always use caution during the first trimester of pregnancy. Breastfeeding note: safe for mom doesn't mean safe for baby to inhale. Use age guidelines when using around your children.

DOG SAFETY

 **SAFE for dogs** when safe guidelines are followed.

 **NOT SAFE for dogs** to inhale, use topically, or consume

Pets note: This is for informational purposes only. This is not a book for using with dogs - but a note for safety when you diffuse or use around your dog. For usage with dogs, refer to my book Using Essential Oils Safely with DOGS. https://www.LeaJacobson.com/DOGBOOK

Essential oils are never recommended to use around cats.

SAFETY ISSUES AFFECTING EVERYONE

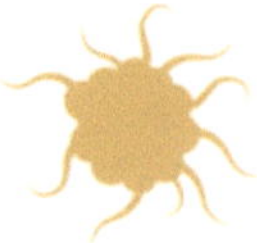

Indicates the essential oil has one or more constituents shown to be **carcinogenic**. This essential oil is best avoided.

Indicates the essential oil contains one or more constituents shown to be **toxic**. This affects everyone, and is best avoided using, especailly if the essential oil has a low topical max.

This essential oil is anti-coagulant, which means it is **blood-thinning** when inhaled, applied topically, or ingested. Can be especially dangerous if taking **aspirin** or **blood thinners**, and before and after **surgery**.

Indicates the essential oil is **phototoxic** when applied to the skin above the topical max dilution in the presence of UV light (sun or tanning beds). If you apply to your skin, dilute to no higher than the topical max to stay safe!

SAFETY ISSUES AFFECTING SOME

Indicates the essential oil has interactions when taking certain **medications**.

Indicates the essential oil has specific **contraindications.**

TOPICAL DILUTION MAXIMUMS

Using more than the amount listed when applying to the skin puts you at risk for adverse reactions such as redness, irritation, phototoxicity, sensitization, etc. If no topical dilution maximum is listed, follow safe dilution guidelines.

A-Z ESSENTIAL OILS

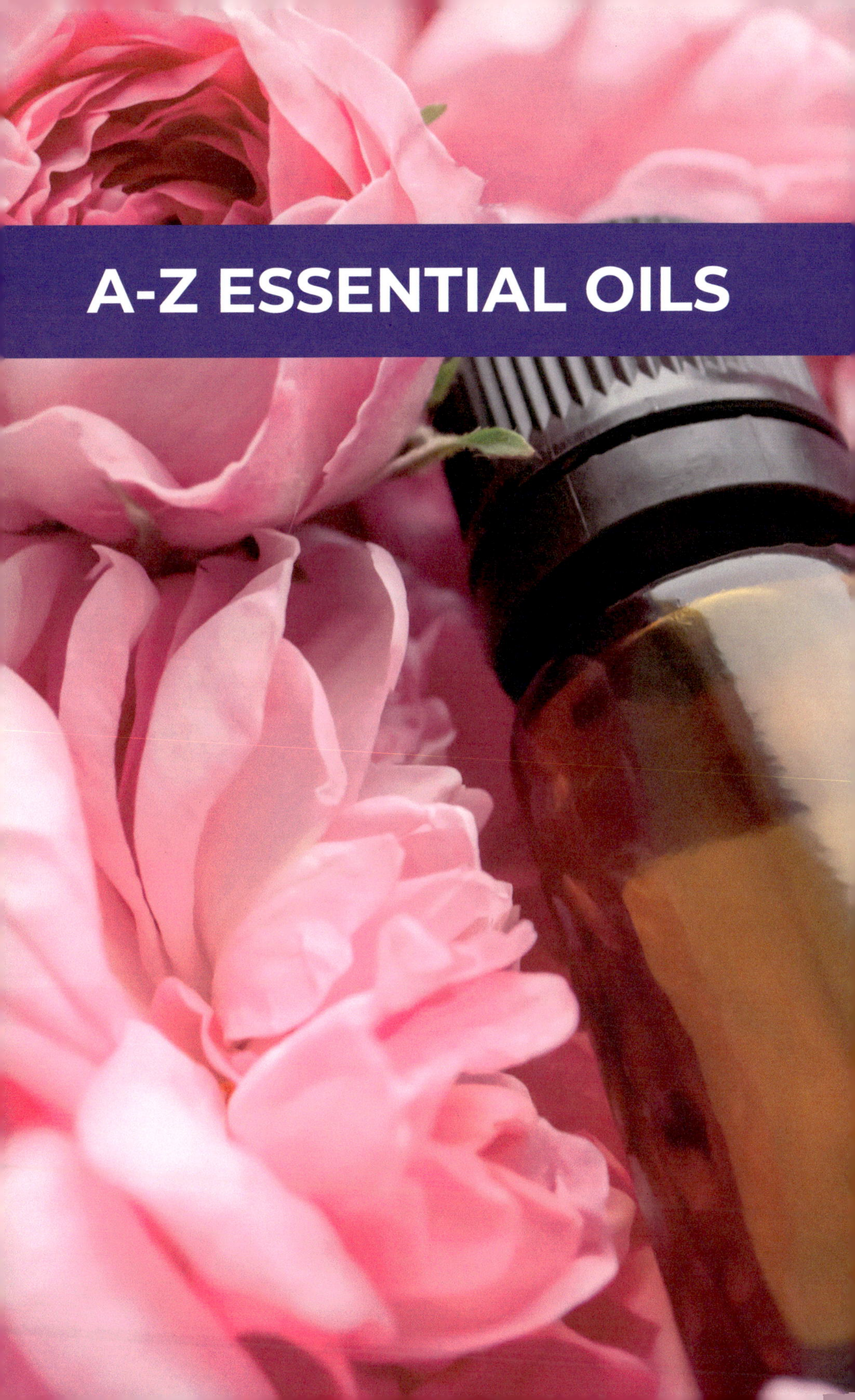

ALLSPICE / PIMENTO BERRY

Pimenta officinalis, Pimenta dioica

DO NOT USE - UNSAFE FOR ALL

This essential oil contains one or more constituents shown to be **carcinogenic** when inhaled, applied topically, or ingested.

This essential oil is anti-coagulant, which means it is **blood-thinning** when inhaled, applied topically, or ingested. Can be especially dangerous if taking **aspirin** or **blood thinners**, and before and after **surgery**.

TOPICAL MAX: 0.15%. For informational purposes only - do not use.

AMYRIS / WEST INDIAN SANDALWOOD

Amyris balsamifera

No topical max - but always dilute before applying to the skin.

ANGELICA ROOT

Angelica archangelica

TOPICAL MAX: 0.8%

This essential oil is **phototoxic** when more than **0.8%** is applied to the skin in the presence of UV light (sun or tanning beds).

ANGELICA ROOT (Himalayan)

Angelica glauca

No topical max - but always dilute before applying to the skin.

ANGELICA SEED

Angelica archangelica

No topical max - but always dilute before applying to the skin.

ANISE

Pimpinella anisum

ANISE (STAR)

Illicium verum

DO NOT USE - UNSAFE FOR ALL

These essential oils contain one or more constituents shown to be **carcinogenic** when inhaled, applied topically, or ingested.

These essential oils are anti-coagulant, which means they are **blood-thinning** when inhaled, applied topically, or ingested. Can be especially dangerous if taking **aspirin** or **blood thinners**, and before and after **surgery**.

These essential oils are a reproductive hormone modulator and should not be inhaled, used topically, or ingested during **hormone replacement therapy, oral contraceptives,** and **progesterone therapy.**

These essential oils have contraindications with **endometriosis** and **estrogen-dependent cancers** when inhaled, applied topically, or ingested.

These essential oils have interactions with **diabetes medication** when used internally (we don't recommend ingesting essential oils).

TOPICAL MAX: 2.4% for Anise and 1.7% for Anise (Star). For informational purposes only - do not use.

ARBORVITAE / THUJA / WESTERN RED CEDAR

Thuja plicata

DO NOT USE - UNSAFE FOR ALL

This essential oil contains one or more constituents shown to be **neurotoxic** when inhaled, applied topically, or ingested.

This essential oil is potentially convulsant and has contraindications with **epilepsy** when inhaled, applied topically, or ingested.

TOPICAL MAX: 0.25%. For informational purposes only - do not use.

BALSAM POPLAR

Populus balsamifera

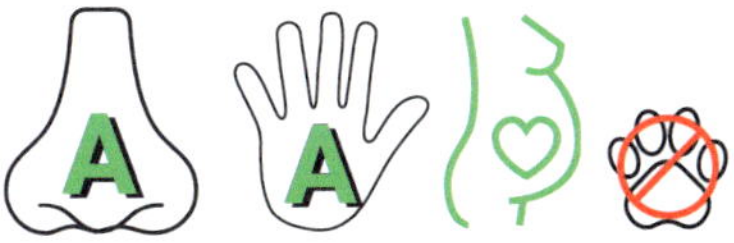

This essential oil has interactions with **drugs metabolized by CYP2D6** when inhaled, applied topically, or ingested.

No topical max - but always dilute before applying to the skin.

BASIL (estragole)

Ocimum basilicum ct estragole

BASIL (Madagascan)

Ocimum gratissimum, Ocimum viride

BASIL (pungent)

Ocimum gratissimum, Ocimum viride

DO NOT USE - UNSAFE FOR ALL

These essential oils contain one or more constituents shown to be **carcinogenic** when inhaled, applied topically, or ingested.

These essential oils are anti-coagulant, which means **blood-thinning** when inhaled, applied topically, or ingested. Can be especially dangerous if taking **aspirin** or **blood thinners**, and before and after **surgery**.

Basil (pungent) essential oil has interactions with **pethidine**, **MAOIs**, and **SSRIs** when used internally (we don't recommend ingesting essential oils).

TOPICAL MAX: 0.1% for Basil (estragole), and 0.2% for Basil (Madagascan), and 0.8% for Basil (pungent). For informational purposes only - do not use.

BASIL (hairy)

Ocimum americanum var pilosum, Ocimum canum

TOPICAL MAX: 30%

BASIL (holy)

Ocimum tenuiflorum, Ocimum sanctum

TOPICAL MAX: 1%

This essential oil is anti-coagulant, which means **blood-thinning** when inhaled, applied topically, or ingested. Can be especially dangerous if taking **aspirin** or **blood thinners**, and before and after **surgery**.

This essential oil has interactions with **pethidine**, **MAOIs**, and **SSRIs** when used internally (we don't recommend ingesting essential oils).

? BASIL (lemon)

Ocimum x citriodorum

TOPICAL MAX: 1.4%

This essential oil has interactions with **diabetes medication**, and **drugs metabolized by CYP2B6** when used internally (we don't recommend ingesting essential oils).

? BASIL (linalool)

Ocimum basilicum ct linalool

TOPICAL MAX: 3.3%

? BASIL (methyl cinnamate)

Ocimum basilicum ct methyl cinnamate

TOPICAL MAX: 15%

BAY (West Indian)

Pimenta racemosa var racemosa, Pimenta acris

TOPICAL MAX: 0.9%

This essential oil is anti-coagulant, which means **blood-thinning** when inhaled, applied topically, or ingested. Can be especially dangerous if taking **aspirin** or **blood thinners**, and before and after **surgery**.

This essential oil has interactions with **pethidine**, **MAOIs**, and **SSRIs** when used internally (we don't recommend ingesting essential oils).

BENZOIN

Styrax benzoin, Styrax paralleloneurus, Styrax tonkinensis

TOPICAL MAX: 2%

BERGAMOT (cold pressed)

Citrus bergamia, Citrus aurantium

TOPICAL MAX: 0.4%

This essential oil is **phototoxic** when more than **0.4%** is applied to the skin in the presence of UV light (sun or tanning beds).

BERGAMOT (steam distilled)

Citrus bergamia, Citrus aurantium

No topical max - but always dilute before applying to the skin.

BERGAMOT MINT

Mentha aquatica var citrata, Mentha citrata

No topical max - but always dilute before applying to the skin.

BIRCH (sweet)

Betula lenta

TOPICAL MAX: 2.5%

This essential oil contains one or more constituents shown to be **moderately toxic**.

This essential oil is anti-coagulant, which means **blood-thinning** when inhaled, applied topically, or ingested. Can be especially dangerous if taking **aspirin** or **blood thinners**, and before and after **surgery**.

This essential oil is potentially convulsant and has contraindications with **epilepsy** when inhaled, applied topically, or ingested.

This essential oil should not be inhaled, applied topically, or ingested with **salicylate sensitivity** and **ADHD**.

This essential oil should be avoided with **GERD** when used internally (we don't recommend ingesting essential oils).

BLACK SEED

Nigella sativa

This essential oil has interactions with **diabetes medication** when used internally (we don't recommend ingesting essential oils).

No topical max - but always dilute before applying to the skin.

BUCHU (diosphenol)

Agathosma betulina, Barosma betulina

BUCHU (pulegone)

Agathosma crenulata, Barosma crenulata

DO NOT USE - UNSAFE FOR ALL

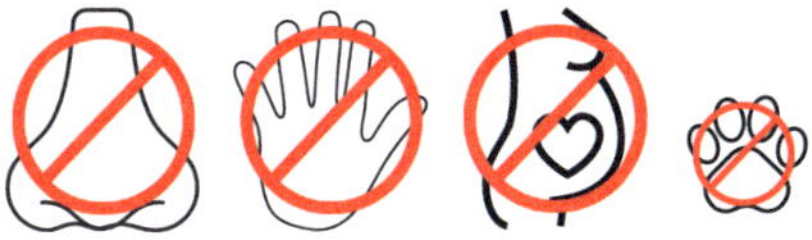

This essential oil contains one or more constituents shown to be **hepatotoxic**.

This essential oil is potentially convulsant and has contraindications with **epilepsy** when inhaled, applied topically, or ingested.

TOPICAL MAX: 11% for Buchu (disophenol), and 1.4% for Buchu (pulegone). For informational purposes only - do not use.

BUDDHA WOOD

Eremophila mitchellii

No topical max - but always dilute before applying to the skin.

CAJEPUT

Melaleuca cajuputi, Melaleuca leucadendron var. cajuputi

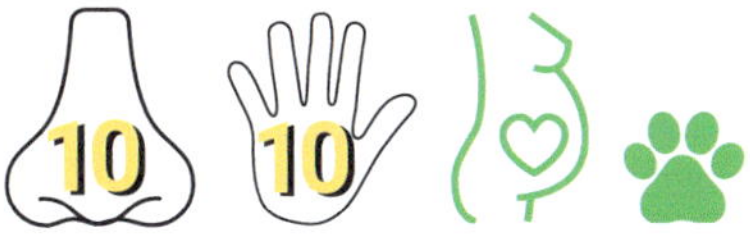

No topical max - but always dilute before applying to the skin.

CAMPHOR (Borneo)

Dryobalanops aromatica, Dryobalanops camphora

CAMPHOR (white)

Cinnamomum camphora, Laurus camphora

No topical max - but always dilute before applying to the skin.

CAMPHOR (brown)

Cinnamomum camphora, Laurus camphora

CAMPHOR (yellow)

Cinnamomum camphora, Laurus camphora

DO NOT USE - UNSAFE FOR ALL

These essential oils contain one or more constituents shown to be **carcinogenic** when inhaled, applied topically, or ingested.

TOPICAL MAX: 0.25% for Camphor (yellow). For informational purposes only - do not use.

CANANGA

Cananga odorata

TOPICAL MAX: 0.8%

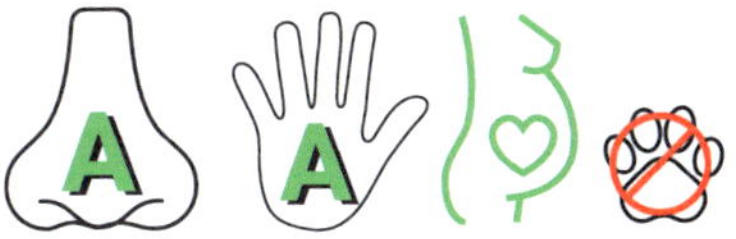

CARAWAY

Carum carvi

No topical max - but always dilute before applying to the skin.

? CARDAMOM

Elettaria cardamomum

? CARDAMOM (black)

Amomum subulatum

No topical max - but always dilute before applying to the skin.

CARROT SEED

Daucus carota

No topical max - but always dilute before applying to the skin.

CASSIA

Cinnamomum cassia, Cinnamomum aromaticum

TOPICAL MAX: 0.05%

This essential oil is anti-coagulant, which means **blood-thinning** when inhaled, applied topically, or ingested. Can be especially dangerous if taking **aspirin** or **blood thinners**, and before and after **surgery**.

This essential oil has interactions with **diabetes medication** when used internally (we don't recommend ingesting essential oils).

CATNIP

Nepeta cataria

This essential oil contains one or more constituents shown to be **psychotropic**.

No topical max - but always dilute before applying to the skin.

CEDARWOOD (Atlas)

Cedrus atlantica

CEDARWOOD (Texan)

Juniperus ashei, Juniperus mexicana

CEDARWOOD (Virginian)

Juniperus virginiana

No topical max - but always dilute before applying to the skin.

CEDARWOOD (Chinese)

Chamaecyparis funebris, Cupressus funebris

CEDARWOOD (Port Orford)

Chamaecyparis lawsoniana

No topical max - but always dilute before applying to the skin.

CEDARWOOD (Himalyan)

Cedrus deodara

May be moderately toxic.

No topical max - but always dilute before applying to the skin.

CHAMOMILE (German)

Matricaria chamomilla, Matricaria recutita, Chamomilla recutita

This essential oil has interactions with **drugs metabolized by CYP2D6** when inhaled, applied topically, or ingested.

This essential oil has interactions with **drugs metabolized by CYP1A2, CYP2C9,** and **CYP3A4** when used internally (we don't recommend ingesting essential oils).

No topical max - but always dilute before applying to the skin.

CHAMOMILE (Moroccan)

Ormenis mixta, Ormenis multicaulis

No topical max - but always dilute before applying to the skin.

CHAMOMILE (Roman)

Anthemis nobilis, Chamaemelum nobile

No topical max - but always dilute before applying to the skin.

CHAMPACA (orange)

Michelia champaca

TOPICAL MAX: 17.5%

CHASTE TREE / VITEX

Vitex agnus castus

DO NOT USE - UNSAFE FOR ALL

These essential oils are a reproductive hormone modulator and should not be inhaled, used topically, or ingested during **hormone replacement therapy, oral contraceptives,** and **progesterone therapy.**

These essential oils have contraindications with **endometriosis** and **estrogen-dependent cancers** when inhaled, applied topically, or ingested.

This essential oil has interactions with **drugs metabolized by CYP2D6** when used internally (we don't recommend ingesting essential oils).

CILANTRO / CORIANDER LEAF

Coriandrum sativum

No topical max - but always dilute before applying to the skin.

CINNAMON BARK

Cinnamomum verum, Cinnamomum zeylanicum

TOPICAL MAX: 0.01%

This essential oil is anti-coagulant, which means **blood-thinning** when inhaled, applied topically, or ingested. Can be especially dangerous if taking **aspirin** or **blood thinners**, and before and after **surgery**.

This essential oil has interactions with **diabetes medication** when used internally (we don't recommend ingesting essential oils).

CINNAMON LEAF

Cinnamomum verum, Cinnamomum zeylanicum

TOPICAL MAX: 0.6%

This essential oil is anti-coagulant, which means **blood-thinning** when inhaled, applied topically, or ingested. Can be especially dangerous if taking **aspirin** or **blood thinners**, and before and after **surgery**.

This essential oil has interactions with **pethidine, MAOIs, SSRIs** when used internally (we don't recommend ingesting essential oils).

CISTUS

Cistus ladanifer, Cistus ladaniferus

No topical max - but always dilute before applying to the skin.

?

CITRONELLA

Cymbopogon winterianus, Cymbopogon nardus

TOPICAL MAX: 18.2%

This essential oil has interactions with **drugs metabolized by CYP2B6** when used internally (we don't recommend ingesting essential oils).

CLARY SAGE

Salvia sclarea

No topical max - but always dilute before applying to the skin.

CLEMENTINE

Citrus clementina, Citrus reticulata var. clementina

No topical max - but always dilute before applying to the skin.

CLOVE

Eugenia caryophyllata, Syzygium aromaticum, Eugenia aromatica

TOPICAL MAX: 0.5%

This essential oil is anti-coagulant, which means **blood-thinning** when inhaled, applied topically, or ingested. Can be especially dangerous if taking **aspirin** or **blood thinners**, and before and after **surgery**.

This essential oil has interactions with **pethidine, MAOIs, SSRIs** when used internally (we don't recommend ingesting essential oils).

COFFEE

Coffea arabica

No topical max - but always dilute before applying to the skin.

COPAIBA

Copaifera officinalis, Copaifera langsdorfii

No topical max - but always dilute before applying to the skin.

CORIANDER

Coriandrum sativum

No topical max - but always dilute before applying to the skin.

COSTUS

Saussurea costus, Aplotaxis lappa, Aucklandia costus, Saussurea lappa

No topical max - but always dilute before applying to the skin.

CUMIN

Cuminum cyminum

TOPICAL MAX: 0.4%

This essential oil is **phototoxic** when more than **0.4%** is applied to the skin in the presence of UV light (sun or tanning beds).

CYPRESS

Cupressus sempervirens

No topical max - but always dilute before applying to the skin.

CYPRESS (blue)

Callitris intratropica

This essential oil has interactions with **drugs metabolized by CYP2D6** when used internally (we don't recommend ingesting essential oils).

No topical max - but always dilute before applying to the skin.

CYPRIOL / NARGAMOTHA

Cyperus scariosus, Cyperus mitis

No topical max - but always dilute before applying to the skin.

DAMIANA

Turnera diffusa, Turnera aphrodisiaca

TOPICAL MAX: 30%

DAVANA

Artemisia pallens

No topical max - but always dilute before applying to the skin.

DILL SEED (European)

Anethum graveolens

This essential oil has interactions with **diabetes medication** when used internally (we don't recommend ingesting essential oils).

No topical max - but always dilute before applying to the skin.

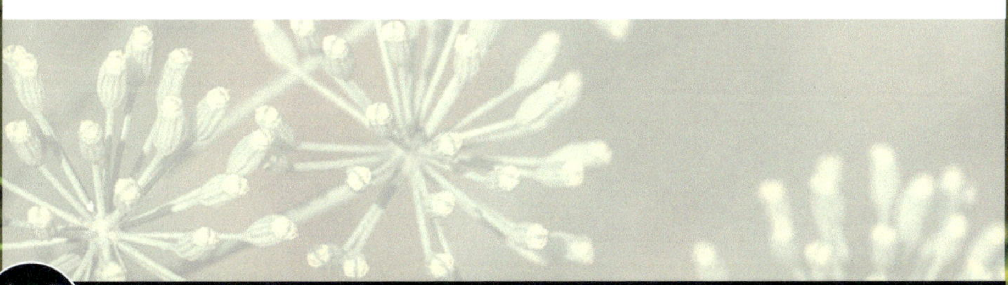

DILL SEED (Indian)

Anethum sowa

DO NOT USE - UNSAFE FOR ALL

This essential oil contains one or more constituents shown to be **hepatotoxic** and **nephrotoxic**.

TOPICAL MAX: 1.4%. For informational purposes only - do not use.

? DILL WEED

Anethym graveolens

This essential oil has interactions with **diabetes medication** when used internally (we don't recommend ingesting essential oils).

No topical max - but always dilute before applying to the skin.

ELEMI

Canarium luzonicum, Canarium vulgare

No topical max - but always dilute before applying to the skin.

EUCALYPTUS (cineole)

Eucalyptus globulus, Eucalyptus radiata, Eualyptus smithii, etc.

TOPICAL MAX: 20%

EUCALYPTUS (macarthurii)

Eucalyptus macarthurii

This essential oil has contraindications with **cholestasis** when used internally (we don't recommend ingesting essential oils).

No topical max - but always dilute before applying to the skin.

EUCALYPTUS (peppermint)

Eucalyptus dives

No topical max - but always dilute before applying to the skin.

EUCALYPTUS (polybractea)

Eucalyptus polybractea ct cryptone

No topical max - but always dilute before applying to the skin.

FENNEL (Bitter)

Foeniculum vulgare

FENNEL (Sweet)

Foeniculum vulgare

DO NOT USE - UNSAFE FOR ALL

These essential oils contain one or more constituents shown to be **carcinogenic** when inhaled, applied topically, or ingested.

These essential oils are anti-coagulant, which means they are **blood-thinning** when inhaled, applied topically, or ingested. Can be especially dangerous if taking **aspirin** or **blood thinners**, and before and after **surgery**.

These essential oils are a reproductive hormone modulator and should not be inhaled, used topically, or ingested during **hormone replacement therapy, oral contraceptives**, and **progesterone therapy.**

These essential oils have contraindications with **endometriosis** and **estrogen-dependent cancers** when inhaled, applied topically, or ingested.

These essential oils have interactions with **diabetes medication** when used internally (we don't recommend ingesting essential oils).

TOPICAL MAX: 1.8% for Bitter and 2.5% for Sweet. For informational purposes only - do not use.

FIR (Douglas)

Pseudotsuga menziesii

FIR CONES (silver)

Abies alba

FIR NEEDLE (Canadian)

Abies balsamea

FIR NEEDLE (Himalayan)

Abies spectabilis, Abies webbiana, Pinus spectabilis, Pinus webbiana

FIR NEEDLE (Japanese)

Abies sachalinensis

FIR NEEDLE (Siberian)

Abies sibirica

FIR NEEDLE (silver)

Abies alba

No topical max - but always dilute before applying to the skin.

FRAGONIA

Agonis fragrans

No topical max - but always dilute before applying to the skin.

FRANKINCENSE

Boswellia carterii/sacra, Boswellia frereana, Boswellia serrata

No topical max - but always dilute before applying to the skin.

FRANKINCENSE (papyrifera)

Boswellia papyrifera

No topical max - but always dilute before applying to the skin.

GALBANUM

Ferula galbaniflua, Ferula gummosa, Ferula erubescens

No topical max - but always dilute before applying to the skin.

GARLIC

Allium sativum

This essential oil is anti-coagulant, which means **blood-thinning** when inhaled, applied topically, or ingested. Can be especially dangerous if taking **aspirin** or **blood thinners**, and before and after **surgery**.

No topical max - but always dilute before applying to the skin.

GERANIUM

Pelargonium graveolens, Pelargonium x asperum

TOPICAL MAX: 17.5%

These essential oils have interactions with **diabetes medication** when used internally (we don't recommend ingesting essential oils).

This essential oil has interactions with **drugs metabolized by CYP2B6** when used internally (we don't recommend ingesting essential oils).

GHANDI ROOT

Homalomena aromatica

No topical max - but always dilute before applying to the skin.

GINGER

Zingiber officinale

No topical max - but always dilute before applying to the skin.

GINGERGRASS

Cymbopogon martinii, Andropogon martinii

No topical max - but always dilute before applying to the skin.

? GRAPEFRUIT (cold pressed)

Citrus paradisi

TOPICAL MAX: 4%

This essential oil is **phototoxic** when more than **4%** is applied to the skin in the presence of UV light (sun or tanning beds).

✓ GRAPEFRUIT (steam distilled)

Citrus paradisi

No topical max - but always dilute before applying to the skin.

GURJUN BALSAM

Dipterocarpus turbinatus, Dipterocarpus tuberculatus

No topical max - but always dilute before applying to the skin.

HELICHRYSUM

Helichrysum italicum/Helichrysum angustifolium, Helchrysum stoechas

No topical max - but always dilute before applying to the skin.

HELICHRYSUM (gymnocephalum)

Helichrysum gymnocephalum

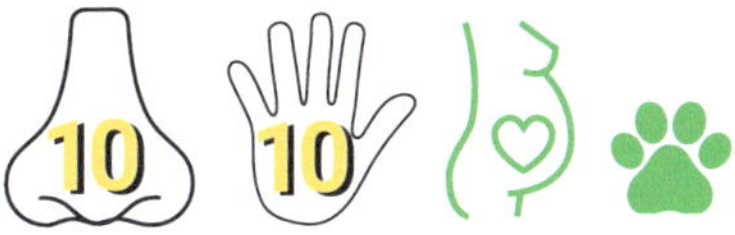

No topical max - but always dilute before applying to the skin.

HEMP

Cannabis sativa, Cannabis indica

No topical max - but always dilute before applying to the skin.

HINOKI

Chamaecyparis obtusa

No topical max - but always dilute before applying to the skin.

HO LEAF (camphor)

Cinnamomum camphora ct camphor

DO NOT USE - UNSAFE FOR ALL

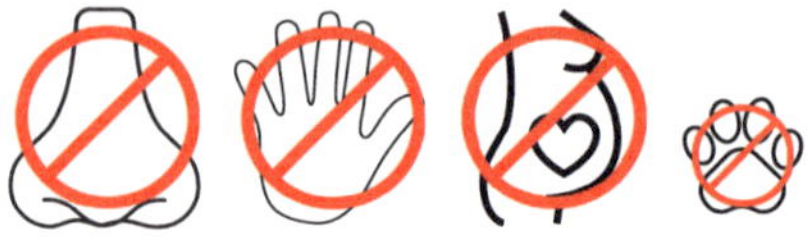

This essential oil contains one or more constituents shown to be **neurotoxic** when inhaled, applied topically, or ingested.

This essential oil is potentially convulsant and has contraindications with **epilepsy** when inhaled, applied topically, or ingested.

This essential oil has interactions with **drugs metabolized by CYP2E1** when used internally (we don't recommend ingesting essential oils).

TOPICAL MAX: 0.8%. For informational purposes only - do not use.

? HO LEAF (cineole) / RAVINTSARA

Cinnamomum camphora ct. cineole

TOPICAL MAX: 11%

✓ HO LEAF (linalool) / HO WOOD

Cinnamomum camphora spp glavescens, *Cinnamomum camphora* ct linalool

No topical max - but always dilute before applying to the skin.

HYSSOP (linalool)

Hyssopus officinalis var decumbens ct linalool

No topical max - but always dilute before applying to the skin.

HYSSOP (pinocamphone)

Hyssopus officinalis ct pinocamphone

DO NOT USE - UNSAFE FOR ALL

This essential oil contains one or more constituents shown to be **carcinogenic** when inhaled, applied topically, or ingested.

This essential oil contains one or more constituents shown to be **neurotoxic** when inhaled, applied topically, or ingested.

This essential oil is potentially convulsant and has contraindications with **epilepsy** when inhaled, applied topically, or ingested.

TOPICAL MAX: 0.3%. For informational purposes only - do not use.

INULA

Dittrichia graveolens, Inula graveolens

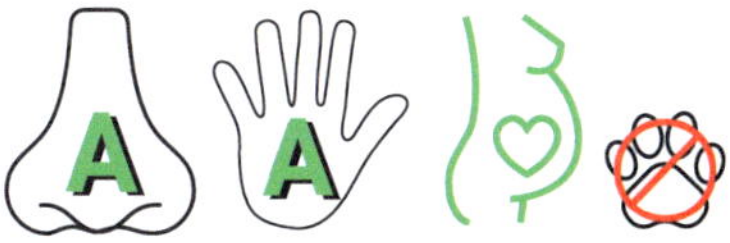

No topical max - but always dilute before applying to the skin.

JASMINE ABSOLUTE

Jasminum grandiflorum, Jasminum officinale

TOPICAL MAX: 0.7%

JASMINE (sambac)

Jasminum sambac

TOPICAL MAX: 4%

This essential oil has interactions with **drugs metabolized by CYP2D6** when used internally (we don't recommend ingesting essential oils).

JUNIPER (Phoenician)

Juniperus phoenicea

No topical max - but always dilute before applying to the skin.

JUNIPER (Utah)

Juniperus osteosperma

No topical max - but always dilute before applying to the skin.

JUNIPERBERRY

Juniperus communis

No topical max - but always dilute before applying to the skin.

KANUKA

Kunzea ericoides

No topical max - but always dilute before applying to the skin.

KUNZEA

Kunzea ambigua

No topical max - but always dilute before applying to the skin.

LABDANUM

Cistus ladanifer

No topical max - but always dilute before applying to the skin.

LAUREL LEAF

Laurus nobilis

DO NOT USE - UNSAFE FOR ALL

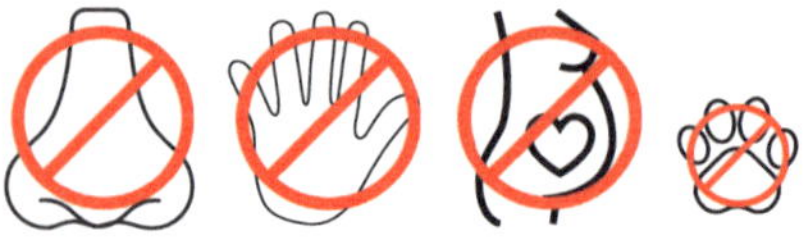

This essential oil contains one or more constituents shown to be **carcinogenic** when inhaled, applied topically, or ingested.

TOPICAL MAX: 0.5%. For informational purposes only - do not use.

LAVANDIN

Lavandula x intermedia, Lavandula hybrida, Lavandula hortensis, Lavandin abrialis, Lavandin grosso, Lavandin super

This essential oil is anti-coagulant, which means **blood-thinning** when inhaled, applied topically, or ingested. Can be especially dangerous if taking **aspirin** or **blood thinners**, and before and after **surgery**.

No topical max - but always dilute before applying to the skin.

LAVANDIN ABSOLUTE

Lavandula x intermedia, Lavandula hybrida, Lavandula hortensis

TOPICAL MAX: 0.03%

LAVENDER (Bulgarian)

Lavandula angustifolia, Lavandula officinalis, Lavandula vera

No topical max - but always dilute before applying to the skin.

LAVENDER (French)

Lavandula angustifolia, Lavandula officinalis, Lavandula vera

No topical max - but always dilute before applying to the skin.

LAVENDER (Spanish)

Lavandula stoechas

TOPICAL MAX: 8%

This essential oil contains one or more constituents shown to be **neurotoxic** when inhaled, applied topically, or ingested.

This essential oil is potentially convulsant and has contraindications with **epilepsy** when inhaled, applied topically, or ingested.

LAVENDER (Spike)

Lavandula latifolia, Lavandula spica

TOPICAL MAX: 19%

This essential oil contains one or more constituents shown to be **neurotoxic** when inhaled, applied topically, or ingested.

This essential oil is potentially convulsant and has contraindications with **epilepsy** when inhaled, applied topically, or ingested.

LEDUM

Ledum groenlandicum

No topical max - but always dilute before applying to the skin.

? LEMON (cold pressed)

Citrus limon

TOPICAL MAX: 2%

This essential oil is **phototoxic** when more than **2%** is applied to the skin in the presence of UV light (sun or tanning beds).

? LEMON (steam distilled)

Citrus limon

TOPICAL MAX: 20%

This essential oil is **phototoxic** when more than **20%** is applied to the skin in the presence of UV light (sun or tanning beds).

LEMON BALM (Australian)

Eucalyptus staigeriana

TOPICAL MAX: 3.4%

This essential oil has interactions **drugs metabolized by CYP2B6** when used internally (we don't recommend ingesting essential oils).

LEMON EUCALYPTUS

Eucalyptus citriodora, Corymbia citriodora

No topical max - but always dilute before applying to the skin.

?

LEMON LEAF

Citrus x limon, Citrus limonum

TOPICAL MAX: 1.2%

This essential oil is **phototoxic** when more than **1.2%** is applied to the skin in the presence of UV light (sun or tanning beds).

These essential oils have interactions with **diabetes medication** when used internally (we don't recommend ingesting essential oils).

This essential oil has interactions with **drugs metabolized by CYP2B6** when used internally (we don't recommend ingesting essential oils).

? LEMONGRASS

Cymbopogon flexuosus, Andropogon flexuosus (East Indian); Cymbopogon citratus, Andropogon citratus (West Indian)

TOPICAL MAX: 0.7%

This essential oil has interactions with **drugs metabolized by CYP2B6** when inhaled, applied topically, or ingested.

These essential oils have interactions with **diabetes medication** when used internally (we don't recommend ingesting essential oils).

LIME (Key)

Citrus aurantifolia, Citrus x aurantifolia

LIME (Mexican)

Citrus aurantifolia, Citrus x aurantifolia

LIME (Persian)

Citrus x latifolia

LIME (cold pressed)

Citrus aurantifolia

TOPICAL MAX: 0.7%

This essential oil is **phototoxic** when more than **0.7%** is applied to the skin in the presence of UV light (sun or tanning beds).

LIME (steam distilled)

Citrus aurantifolia

No topical max - but always dilute before applying to the skin.

LIME (Kaffir) / COMBAVA LEAF

Citrus hystrix

No topical max - but always dilute before applying to the skin.

MACE (East Indian)

Myristica fragrans, Myristica officinalis

MACE (Indian)

Myristica fragrans, Myristica officinalis

DO NOT USE - UNSAFE FOR ALL

This essential oil contains one or more constituents shown to be **carcinogenic** when inhaled, applied topically, or ingested.

This essential oil contains one or more constituents shown to be **psychotropic**.

TOPICAL MAX: 2% for East Indian and 4.1% for Indian. For informational purposes only - do not use.

MAGNOLIA FLOWER, MAGNOLIA LEAF

Michelia alba, Michelia longifolia

No topical max - but always dilute before applying to the skin.

MANDARIN

Citrus reticulata, Citrus nobilis

MANDARIN (green)

Citrus reticulata, Citrus nobilis

No topical max - but always dilute before applying to the skin.

MANUKA

Leptospermum scoparium

No topical max - but always dilute before applying to the skin.

? MARJORAM (Spanish)

Thymus mastichina

No topical max - but always dilute before applying to the skin.

✓ MARJORAM (sweet)

Origanum majorana, Origanum hortensis, Origanum dubium

No topical max - but always dilute before applying to the skin.

MARJORAM (wild) carvacrol

Origanum majorana ct carvacrol, *Origanum hortensis* ct carvacrol, *Origanum dubium*

TOPICAL MAX: 1.2%

These essential oils are anti-coagulant, which means they are **blood-thinning** when inhaled, applied topically, or ingested. Can be especially dangerous if taking **aspirin** or **blood thinners**, and before and after **surgery**.

These essential oils have interactions with **diabetes medication** when used internally (we don't recommend ingesting essential oils).

MARJORAM (wild) linalool

Origanum marjorana ct linalool, *Origanum hortensis* ct linalool

TOPICAL MAX: 4.3%

MASTIC

Pistacia lentiscus

TOPICAL MAX: 20%

MAY CHANG / LITSEA

Litsea cubeba, Litsea citrata, Laurus cubeba

TOPICAL MAX: 0.8%

This essential oil has interactions with **drugs metabolized by CYP2B6** when inhaled, applied topically, or ingested.

These essential oils have interactions with **diabetes medication** when used internally (we don't recommend ingesting essential oils).

? MELISSA / LEMON BALM

Melissa officinalis

TOPICAL MAX: 0.9%

These essential oils have interactions with **diabetes medication** when used internally (we don't recommend ingesting essential oils).

This essential oil has interactions with **drugs metabolized by CYP2B6** when used internally (we don't recommend ingesting essential oils).

MUGWORT (camphor/thujone)

Artemisia vulgaris ct camphor/thujone

MUGWORT (chrysanthenyl acetate)

Artemisia vulgaris ct chrysanthenyl acetate

DO NOT USE - UNSAFE FOR ALL

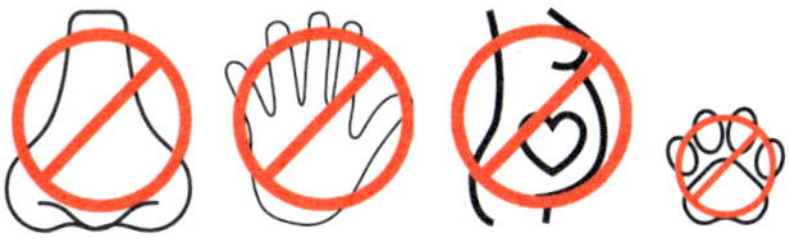

This essential oil contains one or more constituents shown to be **neurotoxic** when inhaled, applied topically, or ingested.

This essential oil is potentially convulsant and has contraindications with **epilepsy** when inhaled, applied topically, or ingested.

TOPICAL MAX: 2% for camphor/thujone, 9.6% for chrysanthenyl acetate. For informational purposes only - do not use.

MUGWORT (Douglas)

Artemisia douglasiana

No topical max - but always dilute before applying to the skin.

MUGWORT (great)

Artemisia arborescens

DO NOT USE - UNSAFE FOR ALL

This essential oil contains one or more constituents shown to be **neurotoxic** when inhaled, applied topically, or ingested.

This essential oil is potentially convulsant and has contraindications with **epilepsy** when inhaled, applied topically, or ingested.

This essential oil has interactions with **drugs metabolized by CYP2D6** when inhaled, applied topically, or ingested.

These essential oils have interactions with **drugs metabolized by CYP1A2** and **CYP3A4** when used internally (we don't recommend ingesting essential oils).

TOPICAL MAX: 0.7%. For informational purposes only - do not use.

MUHUHU

Brachylaena hutchinsii, Brachylaena huillensis

No topical max - but always dilute before applying to the skin.

MUSTARD

Brassica nigra, Brassica juncea

DO NOT USE - UNSAFE FOR ALL

This essential oil contains one or more constituents shown to be **toxic** when inhaled, applied topically, or ingested.

MYRRH

Commiphora myrrha, Commiphora molmol

No topical max - but always dilute before applying to the skin.

MYRTLE

Myrtus communis

DO NOT USE - UNSAFE FOR ALL

This essential oil contains one or more constituents shown to be **carcinogenic** when inhaled, applied topically, or ingested.

These essential oils have interactions with **diabetes medication** when used internally (we don't recommend ingesting essential oils).

TOPICAL MAX: 1.9%. For informational purposes only - do not use.

MYRTLE (honey)

Melaleuca teretifolia

TOPICAL MAX: 0.9%

This essential oil has interactions with **drugs metabolized by CYP2B6** when inhaled, applied topically, or ingested.

These essential oils have interactions with **diabetes medication** when used internally (we don't recommend ingesting essential oils).

MYRTLE (lemon)

Backhousia citriodora

TOPICAL MAX: 0.7%

This essential oil has interactions with **drugs metabolized by CYP2B6** when inhaled, applied topically, or ingested.

These essential oils have interactions with **diabetes medication** when used internally (we don't recommend ingesting essential oils).

NEROLI

Citrus x aurantium

No topical max - but always dilute before applying to the skin.

NIAOULI (cineole)

Melaleuca quinquenervia ct cineole

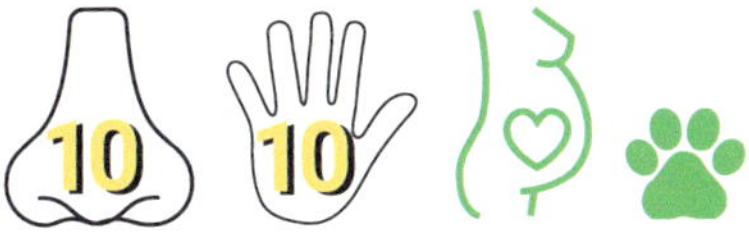

No topical max - but always dilute before applying to the skin.

NIAOULI (nerolidol)

Melaleuca quinquenervia ct nerolidol

NIAOULI (viridiflorol)

Melaleuca quinquenervia ct viridiflorol

NIAOULI / NEROLINA (linalool)

Melaleuca quinquenervia ct linalool

No topical max - but always dilute before applying to the skin.

NOOTKA / CYPRESS LEAF

Cupressus nootkatensis

No topical max - but always dilute before applying to the skin.

NUTMEG (East Indian)

Myristica fragrans, Myristica officinalis, Myristica moschata, Myristica aromatica, Myristica amboinensis

NUTMEG (West Indian)

Myristica fragrans, Myristica officinalis, Myristica moschata, Myristica aromatica, Myristica amboinensis

DO NOT USE - UNSAFE FOR ALL

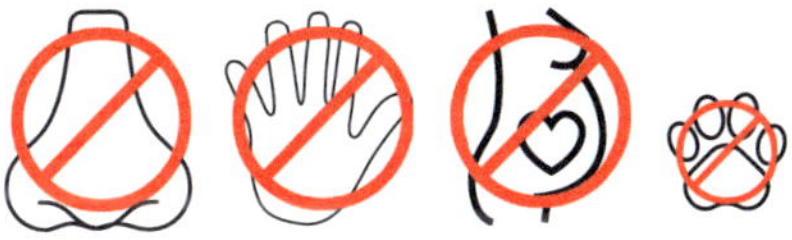

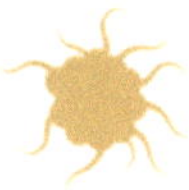

This essential oil contains one or more constituents shown to be **carcinogenic** when inhaled, applied topically, or ingested.

This essential oil contains one or more constituents shown to be **psychotropic**.

TOPICAL MAX: 0.8% for East Indian and 5% for West Indian. For informational purposes only - do not use.

OCOTEA

Ocotea quixos

These essential oils are anti-coagulant, which means they are **blood-thinning** when inhaled, applied topically, or ingested. Can be especially dangerous if taking **aspirin** or **blood thinners**, and before and after **surgery**.

No topical max - but always dilute before applying to the skin.

?

OPOPONAX

Commiphora guidottii

TOPICAL MAX: 0.6%

ORANGE (bitter)

Citrus x aurantium, Citrus aurantium subsp amara

ORANGE (blood)

Citrus sinensis

ORANGE (wild)

Citrus sinensis, Citrus aurantium var sinensis

TOPICAL MAX: 1.25%

This essential oil is **phototoxic** when more than **1.25%** is applied to the skin in the presence of UV light (sun or tanning beds).

ORANGE (flower) ABSOLUTE

Citrus x aurantium

TOPICAL MAX: 1%

ORANGE (sweet)

Citrus sinensis, Citrus aurantium var sinensis

No topical max - but always dilute before applying to the skin.

OREGANO

Origanum vulgare, Origanum onites

TOPICAL MAX: 1.1%

This essential oil is anti-coagulant, which means **blood-thinning** when inhaled, applied topically, or ingested. Can be especially dangerous if taking **aspirin** or **blood thinners**, and before and after **surgery**.

This essential oil has interactions with **diabetes medication** when used internally (we don't recommend ingesting essential oils).

OSMANTHUS ABSOLUTE

Osmanthus fragrans, Olea fragrans

No topical max - but always dilute before applying to the skin.

PALMAROSA

Cymbopogon martinii, Andropogon martinii

TOPICAL MAX: 6.5%

This essential oil has interactions with **drugs metabolized by CYP2B6** when used internally (we don't recommend ingesting essential oils).

PALO SANTO

Bursera graveolens

DO NOT USE - UNSAFE FOR ALL

This essential oil contains one or more constituents shown to be **hepatotoxic**.

TOPICAL MAX: 3.4% For informational purposes only - do not use.

PARSLEY (LEAF)

Petroselinum crispum, Petroselinum sativum

DO NOT USE - UNSAFE FOR ALL

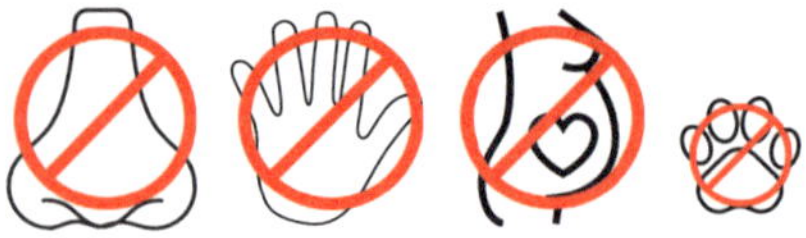

This essential oil contains one or more constituents shown to be **toxic**.

TOPICAL MAX: 14.6%. For informational purposes only - do not use.

PARSLEY (SEED)

Petroselinum crispum, Petroselinum sativum

DO NOT USE - UNSAFE FOR ALL

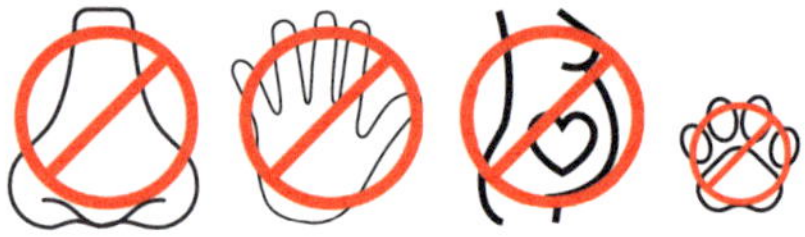

This essential oil contains one or more constituents shown to be **hepatotoxic** and **nephrotoxic**.

This essential oil has interactions with **pethidine**, **MAOIs**, and **SSRIs** when used internally (we don't recommend ingesting essential oils).

TOPICAL MAX: 1.1%. For informational purposes only - do not use.

PATCHOULI

Pogostemon cablin, Pogostemon patchouly

This essential oil is anti-coagulant, which means **blood-thinning** when inhaled, applied topically, or ingested. Can be especially dangerous if taking **aspirin** or **blood thinners**, and before and after **surgery**.

No topical max - but always dilute before applying to the skin.

PENNYROYAL

Hedeoma pulegioides, Mentha pulegium, Micromeria fruticosa

DO NOT USE - UNSAFE FOR ALL

This essential oil contains one or more constituents shown to be **hepatotoxic** and **nephrotoxic**.

This essential oil is potentially convulsant and has contraindications with **epilepsy** when inhaled, applied topically, or ingested.

TOPICAL MAX: 1.3%. For informational purposes only - do not use.

PEPPER (black)

Piper nigrum

No topical max - but always dilute before applying to the skin.

PEPPER (pink)

Schinus molle

No topical max - but always dilute before applying to the skin.

PEPPER (white)

Piper nigrum

No topical max - but always dilute before applying to the skin.

PEPPERMINT

Mentha piperita

TOPICAL MAX: 5.4%

*May cause reduced milk supply in breastfeeding women.

This essential oil should be avoided with **G6PD deficiency** when inhaled, applied topically, or ingested.

This essential oil should be avoided with **cardiac fibrillation** when inhaled, applied topically, or ingested.

This essential oil should be avoided with **GERD** when used internally (we don't recommend ingesting essential oils).

This essential oil has contraindications with **cholestasis** when used internally (we don't recommend ingesting essential oils).

PERU BALSAM

Myroxylon balsamum, Myroxylon pereirae

TOPICAL MAX: 0.4%

PETITGRAIN

Citrus aurantium

No topical max - but always dilute before applying to the skin.

PINE (black)

Pinus nigra

PINE (dwarf)

Pinus mugo, Pinus montana

PINE (grey)

Pinus divaricata, Pinus banksiana

PINE (red)

Pinus resinosa

PINE (Scots)

Pinus sylvestris

PINE (white)

Pinus strobus

No topical max - but always dilute before applying to the skin.

PINE (huon)

Dacrydium franklinii, Lagarostrobos franklinii

PINE (ponderosa)

Pinus ponderosa

DO NOT USE - UNSAFE FOR ALL

This essential oil contains one or more constituents shown to be **carcinogenic** when inhaled, applied topically, or ingested.

TOPICAL MAX: 0.5% for Pine (ponderosa). For informational purposes only - do not use.

PLAI

Zingiber cassumunar, Zingiber montanum

No topical max - but always dilute before applying to the skin.

RAVENSARA BARK

Ravensara aromatica, Ravensara anisata

DO NOT USE - UNSAFE FOR ALL

This essential oil contains one or more constituents shown to be **carcinogenic** when inhaled, applied topically, or ingested.

This essential oil is anti-coagulant, which means **blood-thinning** when inhaled, applied topically, or ingested. Can be especially dangerous if taking **aspirin** or **blood thinners**, and before and after **surgery**.

TOPICAL MAX: 0.1%. For informational purposes only - do not use.

RAVENSARA LEAF

Ravensara aromatica, Ravensara anisata

DO NOT USE - UNSAFE FOR ALL

This essential oil contains one or more constituents shown to be **carcinogenic** when inhaled, applied topically, or ingested.

TOPICAL MAX: 1%. For informational purposes only - do not use.

ROSALINA

Melaleuca ericifolia

No topical max - but always dilute before applying to the skin.

ROSE

Rosa x damascena, Rosa damascena

TOPICAL MAX: 0.6%

ROSE (PROVINCE) ABSOLUTE

Rosa x centifolia, Rosa gallica

TOPICAL MAX: 2.5%

ROSEMARY (a-pinene)

Rosmarinus officinalis ct a-pinene

TOPICAL MAX: 22%

ROSEMARY (camphor)

Rosmarinus officinalis ct camphor

TOPICAL MAX: 16.5%

This essential oil contains one or more constituents shown to be **neurotoxic** when inhaled, applied topically, or ingested.

This essential oil is potentially convulsant and has contraindications with **epilepsy** when inhaled, applied topically, or ingested.

ROSEMARY (cineole)

Rosmarinus officinalis ct cineole

No topical max - but always dilute before applying to the skin.

ROSEMARY (verbenone)

Rosmarinus officinalis ct verbenone

TOPICAL MAX: 6.5

ROSEWOOD

Aniba rosaeodora, Aniba amazonica, Aniba parviflora

No topical max - but always dilute before applying to the skin.

RUE

Ruta graveolens, Ruta montana

TOPICAL MAX: 0.15%

This essential oil is **phototoxic** when more than **0.15%** is applied to the skin in the presence of UV light (sun or tanning beds).

SAGE (African Wild)

Tarchonanthus camphoratus

No topical max - but always dilute before applying to the skin.

SAGE (blue mountain)

Salvia stenophylla

No topical max - but always dilute before applying to the skin.

SAGE (Dalmatian)

Salvia officinalis

DO NOT USE - UNSAFE FOR ALL

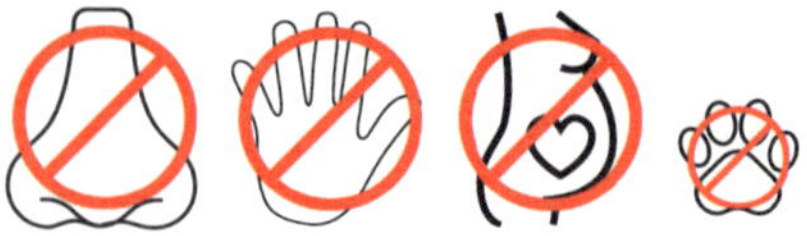

This essential oil contains one or more constituents shown to be **neurotoxic**.

This essential oil is potentially convulsant and has contraindications with **epilepsy** when inhaled, applied topically, or ingested.

TOPICAL MAX: 0.4%. For informational purposes only - do not use.

SAGE (Greek)

Salvia fruticosa, Salvia trilobal

SAGE (white)

Salvia apiana

No topical max - but always dilute before applying to the skin.

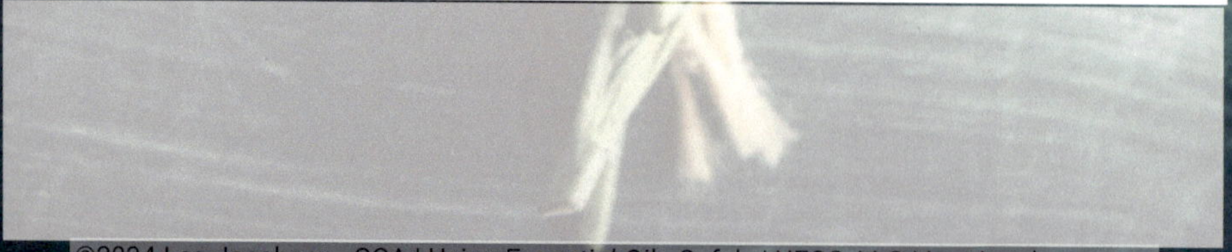

SAGE (Spanish)

Salvia lavandulifolia, Salvia hispanorum

TOPICAL MAX: 12.5%

This essential oil is potentially convulsant and has contraindications with **epilepsy** when inhaled, applied topically, or ingested.

SAGE (wild mountain)

Hemizygia petiolata

TOPICAL MAX: 0.7%

This essential oil has interactions with **drugs metabolized by CYP2D6** when inhaled, applied topically, or ingested.

SANDALWOOD (Australian)

Santalum spicatum, Santalum cygnorum, Fusanus spicatus, Eucarya spicata

This essential oil has interactions with **drugs metabolized by CYP2D6** when inhaled, applied topically, or ingested.

No topical max - but always dilute before applying to the skin.

SANDALWOOD (East African)

Osyris lanceolata

No topical max - but always dilute before applying to the skin.

SANDALWOOD (East Indian)

Santalum album

TOPICAL MAX: 2%

? SANDALWOOD (New Caledonian)

Santalum austrocaledonicum

TOPICAL MAX: 2%

✓ SANDALWOOD (paniculatum)

Santalum paniculatum

No topical max - but always dilute before applying to the skin.

SARO

Cinnamosma fragrans

No topical max - but always dilute before applying to the skin.

SAVORY (summer)

Satureia hortensis

TOPICAL MAX: 1.4%

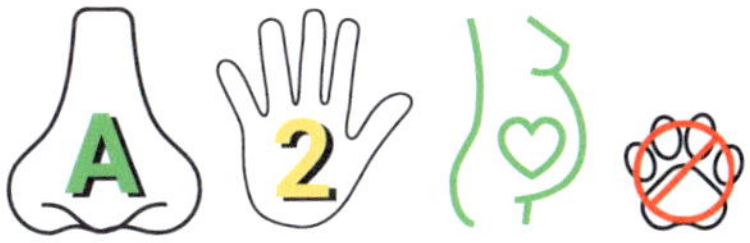

This essential oil is anti-coagulant, which means **blood-thinning** when inhaled, applied topically, or ingested. Can be especially dangerous if taking **aspirin** or **blood thinners**, and before and after **surgery**.

This essential oil has interactions with **diabetes medication** when used internally (we don't recommend ingesting essential oils).

SAVORY (winter)

Satureia montana

TOPICAL MAX: 1.2%

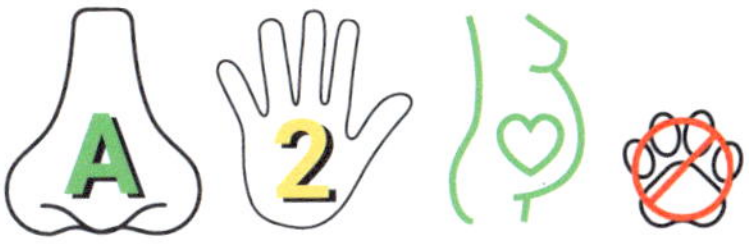

This essential oil is anti-coagulant, which means **blood-thinning** when inhaled, applied topically, or ingested. Can be especially dangerous if taking **aspirin** or **blood thinners**, and before and after **surgery**.

This essential oil has interactions with **diabetes medication** when used internally (we don't recommend ingesting essential oils).

SPEARMINT

Mentha spicata, Mentha cardiaca, Mantha crispa, Mentha viridis

TOPICAL MAX: 1.7%

SPIKENARD

Nardostachys grandiflora

No topical max - but always dilute before applying to the skin.

SPRUCE (black)

Picea mariana, Picea nigra

SPRUCE (Blue Idaho)

Picea pungens

SPRUCE (hemlock)

Tsuga canadensis, Pinus canadensis

SPRUCE (Norway)

Picea abies

SPRUCE (red)

Picea rubens

SPRUCE (white)

Picea glauca, Picea alba

No topical max - but always dilute before applying to the skin.

SUGANDHA

Cinnamomum cecidodaphne, Cinnamomum glaucescens

DO NOT USE - UNSAFE FOR ALL

This essential oil contains one or more constituents shown to be **carcinogenic** when inhaled, applied topically, or ingested.

TOPICAL MAX: 3.3%. For informational purposes only - do not use.

TANGERINE

Citrus reticulata, Citrus nobilis, Citrus tangerine

No topical max - but always dilute before applying to the skin.

TANSY

Tanacetum vulgare, Chrysanthemum tanacetum

DO NOT USE - UNSAFE FOR ALL

This essential oil contains one or more constituents shown to be **neurotoxic**.

This essential oil is potentially convulsant and has contraindications with **epilepsy** when inhaled, applied topically, or ingested.

TOPICAL MAX: 0.5%. For informational purposes only - do not use.

TANSY (Blue)

Tanacetum anuum

This essential oil has interactions with **drugs metabolized by CYP2D6** when inhaled, applied topically, or ingested.

This essential oil has interactions with **drugs metabolized by CYP1A2** and **CYP3A4** when used internally (we don't recommend ingesting essential oils).

No topical max - but always dilute before applying to the skin.

TARRAGON

Artemisia dracunculus

DO NOT USE - UNSAFE FOR ALL

This essential oil contains one or more constituents shown to be **carcinogenic** when inhaled, applied topically, or ingested.

This essential oil is anti-coagulant, which means **blood-thinning** when inhaled, applied topically, or ingested. Can be especially dangerous if taking **aspirin** or **blood thinners**, and before and after **surgery**.

TOPICAL MAX: 0.1%. For informational purposes only - do not use.

TEA TREE

Melaleuca alternifolia

TOPICAL MAX: 15%

TEA TREE (black)

Melaleuca bracteata

DO NOT USE - UNSAFE FOR ALL

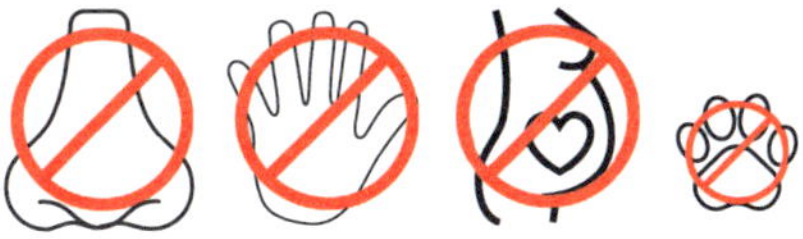

This essential oil contains one or more constituents shown to be **carcinogenic** when inhaled, applied topically, or ingested.

TEA TREE (lemon)

Leptospermum petersonii, Leptospermum citratum

TOPICAL MAX: 0.8%

This essential oil has interactions with **drugs metabolized by CYP2D6** when inhaled, applied topically, or ingested.

This essential oil has interactions with **diabetes medication** when used internally (we don't recommend ingesting essential oils).

THYME (borneol)

Thymus satureioides

TOPICAL MAX: 3.3%

This essential oil is anti-coagulant, which means **blood-thinning** when inhaled, applied topically, or ingested. Can be especially dangerous if taking **aspirin** or **blood thinners**, and before and after **surgery**.

THYME (carvacrol)

Thymus vulgaris ct carvacrol

TOPICAL MAX: 1.3%

This essential oil is anti-coagulant, which means **blood-thinning** when inhaled, applied topically, or ingested. Can be especially dangerous if taking **aspirin** or **blood thinners**, and before and after **surgery**.

? THYME (geraniol)

Thymus vulgaris ct geraniol

This essential oil has interactions with **drugs metabolized by CYP2B6** when used internally (we don't recommend ingesting essential oils).

No topical max - but always dilute before applying to the skin.

? THYME (lemon)

Thymus x citriodorus, Thymus lanuginosus, Thymus serpylum

TOPICAL MAX: 3.7%

This essential oil has interactions with **drugs metabolized by CYP2B6** when used internally (we don't recommend ingesting essential oils).

THYME (limonene)

Thymus vulgaris ct limonene, *Thymus serpyllum* ct limonene

TOPICAL MAX: 2.1%

This essential oil is anti-coagulant, which means **blood-thinning** when inhaled, applied topically, or ingested. Can be especially dangerous if taking **aspirin** or **blood thinners**, and before and after **surgery**.

THYME (linalool)

Thymus vulgaris ct linalool, *Thymus zygis* ct linalool

No topical max - but always dilute before applying to the skin.

THYME (serpyllum)

Thymus serpyllum

TOPICAL MAX: 1.3%

This essential oil is anti-coagulant, which means **blood-thinning** when inhaled, applied topically, or ingested. Can be especially dangerous if taking **aspirin** or **blood thinners**, and before and after **surgery**.

THYME (spike)

Thymbra spicata

TOPICAL MAX: 1.4%

This essential oil is anti-coagulant, which means **blood-thinning** when inhaled, applied topically, or ingested. Can be especially dangerous if taking **aspirin** or **blood thinners**, and before and after **surgery**.

This essential oil has interactions with **diabetes medication** when used internally (we don't recommend ingesting essential oils).

THYME (thujanol)

Thymus vulgaris ct thujanol

No topical max - but always dilute before applying to the skin.

THYME (thymol)

Thymus vulgaris ct thymol, *Thymus zygis* ct thymol

THYME (thymol/carvacrol)

Thymus serpyllum, Thymus zygis

THYME (zygis)

Thymus zygis

TOPICAL MAX: 1.3%

This essential oil is anti-coagulant, which means **blood-thinning** when inhaled, applied topically, or ingested. Can be especially dangerous if taking **aspirin** or **blood thinners**, and before and after **surgery**.

TURMERIC (wild)

Curcuma aromatica

No topical max - but always dilute before applying to the skin.

TURMERIC LEAF

Curcuma longa

No topical max - but always dilute before applying to the skin.

TURMERIC RHIZOME

Curcuma longa, Curcuma domestica

This essential oil has interactions with **diabetes medication** when used internally (we don't recommend ingesting essential oils).

No topical max - but always dilute before applying to the skin.

VALERIAN (European)

Valeriana officinalis, Valeriana fauriei

No topical max - but always dilute before applying to the skin.

VALERIAN (Indian)

Valeriana jatamansi, Nardostachys jatamansi

No topical max - but always dilute before applying to the skin.

VANILLA ABSOLUTE / OLEORESIN

Vanilla planifolia (Bourbon), Vanilla fragrans, Vanilla tahitensis (Tahitian)

No topical max - but always dilute before applying to the skin.

VERBENA (honey)

Lippia javanica

No topical max - but always dilute before applying to the skin.

VERBENA (white)

Lippia alba, Lippia geminata

No topical max - but always dilute before applying to the skin.

VERBENA (lemon)

Aloysia triphylla, Aloysia citriodora

TOPICAL MAX: 0.9%

This essential oil has interactions with **diabetes medication** when used internally (we don't recommend ingesting essential oils).

This essential oil has interactions with **drugs metabolized by CYP2B6** when used internally (we don't recommend ingesting essential oils).

VERBENA (lemon) ABSOLUTE

Aloysia triphylla, Aloysia citriodora

TOPICAL MAX: 0.2%

VETIVER

Vetiveria zizanoides, Andropogon muricatus/zizanoides, Chrysopogon zizanoides, Phalaris zizanoides

TOPICAL MAX: 15%

WINTERGREEN

Gaultheria procumbens, Gaultheria fragrantissima

TOPICAL MAX: 2.4%

This essential oil contains one or more constituents shown to be **toxic**.

This essential oil is anti-coagulant, which means **blood-thinning** when inhaled, applied topically, or ingested. Can be especially dangerous if taking **aspirin** or **blood thinners**, and before and after **surgery**.

This essential oil is potentially convulsant and has contraindications with **epilepsy** when inhaled, applied topically, or ingested.

This essential oil should not be inhaled, applied topically, or ingested with **salicylate sensitivity** and **ADHD**.

This essential oil should be avoided with **GERD** when used internally (we don't recommend ingesting essential oils).

WORMWOOD

Artemisia absinthium

DO NOT USE - UNSAFE FOR ALL

This essential oil contains one or more constituents shown to be **neurotoxic**.

XANTHOXYLUM / ZANTHOXYLUM / TOMAR SEED

Zanthoxylum armatum, Zanthoxylum alatum

No topical max - but always dilute before applying to the skin.

YARROW

Achillea millefolium

TOPICAL MAX: 8.6%

This essential oil contains one or more constituents shown to be **moderately neurotoxic**.

This essential oil is potentially convulsant and has contraindications with **epilepsy** when inhaled, applied topically, or ingested.

This essential oil has interactions with **drugs metabolized by CYP2D6** when inhaled, applied topically, or ingested.

This essential oil has interactions with **drugs metabolized by CYP1A2** and **drugs metabolized by CYP3A4** when used internally (we don't recommend ingesting essential oils).

YARROW (green)

Achillea nobilis, Achilea ligustica

No topical max - but always dilute before applying to the skin.

YLANG YLANG (complete)
YLANG YLANG Extra
YLANG YLANG I
YLANG YLANG II
YLANG YLANG III

Cananga odorata, Cananga odorata genuina

TOPICAL MAX: 0.8%

YUZU

Citrus junos

No topical max - but always dilute before applying to the skin.

FURTHER EDUCATION

Other resources of mine you may enjoy:

- ***Using Essential Oils Safely BOOK*** covers everything from using essential oils with children, to inhalation and topical use best practices.
- **Using Essential Oils Safely COURSE** dives deeper than the book, with an option to earn a certificate! Special membership area allows you to ask Lea questions directly.
- ***Essential Oil Profiles*** covers the safety (ages, "stages" such as pregnancy, drug interactions, pet safety, etc.) and benefits broken down by body system for 60 of the most popular essential oils. Bonus section including dilution charts and several other topics are included.
- ***Essential Oil BLENDS Safety Files*** is a compilation of my thorough safety break-downs for around 800 blends from over 45 companies. You'll see what age(s) the blends are safe for, if they are safe during pregnancy, any drug interactions, and more.
- ***Using Essential Oils Safely with DOGS*** focuses on safety with dogs and other pets
- **Recipes Lab™** is where it's at! Hundreds of recipes safe and effective for the whole family, weekly classes, and so much more.
- **Pocket Aromatherapist™** combines safety + therapeutic properties + benefits in a searchable tool.
- **BRAND REVIEWS Portal** for honest essential oil brand reviews from a Certified Clinical Aromatherapist who doesn't sell essential oils (Lea Jacobson, CCA!).

Be aware of safety issues.
Enjoy all benefits.
Have Peace of mind.
FREE classes: www.LeaJacobson.com

Made in the USA
Las Vegas, NV
07 November 2026

57655928R00107